Y0-DDR-059

by Claire Daniel

SCHOOL PUBLISHERS

Cover, Detlev van Ravenswaay/Photo Researchers, Inc.; 3, Courtesy of NASA; 4, Courtesy NASA/JPL-Caltech; 5, Courtesy of NASA/JPL-Caltech; 6, Courtesy of NASA/JPL-Caltech; 7, Courtesy of NASA; 8, Courtesy of NASA; 9, Courtesy of NASA/JPL-Caltech; 10, Courtesy of NASA; 11, Courtesy of NASA/JPL-Caltech; 12, Courtesy of NASA/JPL-Caltech; 13, Detlev van Ravenswaay/Photo Researchers, Inc.; 14, Navin Patel.

Printed in the United States of America

ISBN 10: 0-15-350593-1
ISBN 13: 978-0-15-350593-5

Ordering Options
ISBN 10: 0-15-350336-X (Grade 6 Below-Level Collection)
ISBN 13: 978-0-15-350336-8 (Grade 6 Below-Level Collection)
ISBN 10: 0-15-357735-5 (package of 5)
ISBN 13: 978-0-15-357735-2 (package of 5)

2 3 4 5 6 7 8 9 10   179   12 11 10 09 08 07

Ever since humans first looked up to the stars, they have wondered about space. With the help of space travel, landing on the moon, and new technology, we have learned much in the past century about the sun and the planets in our solar system.

The Sun

Earth would be a dismal and barren planet without the warmth of the sun. The sun is an average of 93 million miles (150 million km) away from Earth, yet it sends enough heat to allow plants to grow and animals to thrive.

The sun is huge and has a diameter of almost 900,000 miles (1,448,370 km). The 27,000,000° Fahrenheit (14,999,982°C) temperature at its core assures us that enough heat will reach Earth to allow life to flourish.

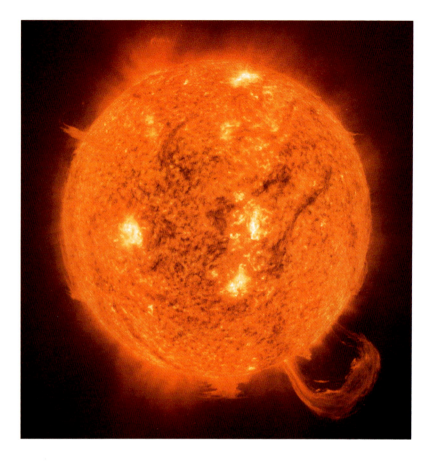

The sun is made up of mostly helium and hydrogen. These gases undergo reactions that release massive amounts of energy. Scientists think that the sun will burn for another five billion years, and that as it does, it will get bigger and hotter. Then after many more millions of years, scientists think that the sun will run out of fuel and become a tiny, cold object known as a "white dwarf."

Mercury

Mercury is the planet that is closest to the sun. That makes it a very hot place—during the day it reaches around 800° Fahrenheit (427°C). However, when the sun disappears, Mercury becomes very cold, around -300° Fahrenheit (-184°C)!

Scientists know that Mercury's surface has been dented by the impact of large rocks in space. The results are huge craters on its surface. The largest chasm on Mercury is quite wide, about 800 miles (1,287 km) across! The surface also looks wrinkled. Scientists think that the wrinkles were created after the planet was formed. They believe that Mercury was once hot, and then it cooled. As it cooled, it shrank, and the surface became wrinkly like a prune.

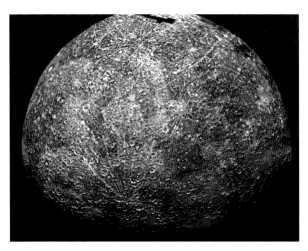

In 1973, a space probe took photographs of Mercury.

Venus

Even though
Venus is the second
closest planet to the
sun, it is actually the
hottest. Thick clouds of
carbon dioxide cover the planet,
trapping heat from the sun. During the
day, temperatures can reach close to
900° Fahrenheit (482°C)!

Venus would make a terrible home for
humans. Hurricane-strength winds blow
constantly. It even rains sulfuric acid on
Venus. Obviously, no plant or animal
found on Earth could survive living on
this harsh planet!

The bluish color of Earth
is made from the oceans of
water that reflect sunlight.

Earth

Earth is the only planet we know of where life has come to thrive. Living things use oxygen in the atmosphere for breathing. There is plenty of water for plants and animals. The sun provides the energy for plant photosynthesis, the basis for the food chain.

From outer space, Earth is distinctive with its blue and green color, with clouds swirling above its surface. We have relatively mild temperatures with a mean temperature of around 59° Fahrenheit (15°C).

Earth's surface is in a state of constant motion. Solid rock plates float on top of molten rock. When these plates hit one another, they change the structure of Earth in different ways. The impact can cause mountain ranges to form over thousands of years. Other times the plates collide to cause earthquakes. Earthquakes can also cause tidal waves.

Mars

Science fiction writers have written many books about martians, but there is no proof that any life on Mars exists. Mars is a much colder planet than Earth. The average temperature on Mars is -67° Fahrenheit (-55°C). Nighttime is much colder than daytime.

Prominent mountains and deep valleys hint that volcanoes were once active on Mars. Not much liquid is found on Mars today, but scientists think that Mars might have once had rivers. There is ice on Mars, but it is made of frozen carbon dioxide, or "dry ice."

The surface of Mars looks red and mottled, and people often call it the "Red Planet." The surface is red because its dust contains iron oxide, or rust. Robotic rovers sent by NASA have been used to travel the surface of Mars to discover more about it.

In 2003, unmanned rovers landed and began exploring the surface of Mars.

Jupiter

Jupiter is the largest planet in our solar system, even though most of it is gas. Jupiter's outer layers are made up of hydrogen gas. At its core, the hydrogen becomes a hard material more like metal. Some scientists think that the temperature inside Jupiter could be hotter than the sun. Clouds engulf Jupiter and swirl around this planet as fast as 250 miles (402 km) per hour.

Jupiter is not a lonely planet. It has sixty-three known moons orbiting it, and there could be even more that have not been discovered! The moon Io is covered with volcanoes, and some are active. Jupiter's moon Europa is made up of ice. Some scientists think there might be oceans of water underneath the ice.

There is a Great Red Spot on Jupiter, which is a hurricane of gases whirling about 22,000 miles per hour.

The spacecraft *Voyager* photographed Saturn's rings.

Saturn

Saturn is the sixth planet from the sun and the second largest planet. Lakes of liquid hydrogen gas cover it's surface. In fact, Saturn is made up entirely of gases except for the rings. They are made up of chunks of ice that stretch miles across. The rings look like only a few bands, but they are actually thousands of separate rings. These icy fragments might be what is left of a small moon that once orbited Saturn.

Saturn has many moons. Some of the moons orbit inside the rings. Titan is the most famous moon. Scientists think that Titan also has liquid gas on its surface.

Uranus

Uranus is also made up entirely of gases. It also has violent winds. It is the third largest planet, and it is very cold. It has an average temperature of -320° Fahrenheit (-196°C). Uranus is about 1.5 billion miles (about 2.5 billion km) away from Earth. It has twenty-seven moons orbiting it.

Uranus also has rings, but they are positioned vertically. This blue-green planet orbits the sun only once every eighty-four years.

William Herschel, an amateur astronomer, discovered Uranus.

Neptune

Like Saturn, Jupiter, and Uranus, Neptune is also made up of gases. Neptune is the eighth and farthest planet from the sun.

Neptune is a brilliant blue color, and it is a very cold and windy planet. Winds can blow up to 1,400 miles (2,253 km) per hour on the surface.

Neptune has a very unusual moon called Triton that orbits Neptune in the direction opposite its other twelve moons! Scientists don't know why this is so. Some believe that the moon might have bumped another body in space and that the impact may have sent the moon orbiting the other way.

Dwarf Planets

Astronomers have long known that our solar system contains many more objects than just the eight planets. Scientists are still deciding what to call some of these objects. In 2006, they decided to call three of these objects "dwarf planets." A dwarf planet looks like a planet. It can even have moons like a planet. However, it is smaller than a planet.

The names of the three dwarf planets are Ceres, Pluto, and Eris. The dwarf planet closest to the sun is Ceres. The next dwarf planet is Pluto. Pluto was considered a planet for many years. In 2006, scientists decided that Pluto was a dwarf planet instead. Pluto has one moon, Charon. The dwarf planet farthest from the sun is Eris.

As scientists learn more about the other objects in our solar system, they may decide to make even more categories for the objects. They could even decide to reclassify existing ones again.

Planet (in order from the sun)	Number of Moons	Mean Temperature °Fahrenheit (°C)	Days to Orbit Sun	Distance from Sun in millions of miles (millions of km)
Mercury	0	333 (167)	88	36 (58)
Venus	0	867 (464)	224.7	67.2 (108.1)
Earth	1	59 (15)	365.2	93 (150)
Mars	2	-85 (-65)	687	141.6 (227.9)
Jupiter	63	-166 (-110)	4,331	483.8 (778.6)
Saturn	47	-220 (-140)	10,747	890.8 (1,433.6)
Uranus	27	-320 (-196)	30,589	2,793.1 (4,495)
Neptune	13	-330 (-201)	59,800	1,784.8 (2,872)

Source: NASA Planetary Fact Sheet